BLUE COLLAR
CHRONICLES

BLUE COLLAR
CHRONICLES

JOHN PROCACCINO

authorHOUSE®

AuthorHouse™ LLC
1663 Liberty Drive
Bloomington, IN 47403
www.authorhouse.com
Phone: 1-800-839-8640

Published by AuthorHouse 11/26/2013

ISBN: 978-1-4918-3900-3 (sc)
ISBN: 978-1-4918-3905-8 (e)

Library of Congress Control Number: 2013921791

CONTENTS

PREFACE

The following may not be enjoyed by certain tradesman. Whether white collar, blue collar, or no collar, this book will offend some readers. Still, it is a truthful account of life in the construction world, particularly in New Jersey. I'm sure anyone in the construction industry, from management to lowly apprentices, will relate to most descriptions in this book. Get ready to be offended.

Although I may poke fun at certain people and situations that I encountered in my forty-plus years in construction, it doesn't mean I don't respect them. If I could do it all over again, I would, because it would mean I could work with so many great people again.

INTRODUCTION

I began my construction career in 1970 as an apprentice in the Trenton, New Jersey, local union for electricians (I.B.E.W. L.U. 269—International Brotherhood of Electrical Workers). I've seen it all— I've been an apprentice, journeyman, shop steward, foreman, general foreman, and project manager.

For over forty years I was fortunate enough to work out of one of the best locals in the country, and work was plentiful and rewarding during most of my years. Between the steel mills; the city's continuing growth; local industries; the number of schools, universities, and colleges; powerhouses; and energy plants, our over 1,100 members stay busy. The world-famous Lower Trenton Bridge, which crosses the mighty Delaware River, says it all in big bright lights: Trenton makes, the world takes.

I was born in Princeton, New Jersey and attended the Princeton school systems from kindergarten through twelfth grade. After high school graduation in 1970, I entered the apprentice program at L.U. 269. This turned out to be a good decision. A week later I got a call from the union hall informing me where to go for my first union job. It was a bit out of character for me to join a powerful union. Everyone who knew me considered me to be timid—and to tell the truth I was more than a little nervous—but I survived.

Unions are scary to a lot of people, but that's because they don't know enough about them. I won't get into the whole controversy about the organized labor movement, but I will say that the basic premise of a union is to level the playing field between management and labor. Think about it: if there weren't unions fighting for better wages and safer conditions, the contractors would be filthy rich (although they are anyway), and there would be sweatshops on every corner. Like any complex issue, there are pros and cons to unions, but in my opinion unions are good for the working man. Union workers are the backbone of the country—they build America. They are blue-collar and middle-class taxpayers. Don't ever underestimate the role of the layman.

THE CONSTRUCTION SITE

Let me begin by describing a typical commercial or industrial construction site in New Jersey, after all of the preliminary planning, town meetings, and architectural work. I'm talking about the laborious work on site.

The construction site is a very busy place, like a small, growing town. It has its own governing body: the mayor is the supervisor, his assistants are the police force, and the construction managers are the judge and jury. To be clear, all these guys are back-stabbing fuck-faces, and not one of them is worth a shit. The tradesmen, carpenters, plumbers, electricians, and so forth, are also the rest of the townspeople.

Once a week there is a town meeting, or job meeting, in which the foreman of each trade faces the mayor, police, judge, and jury. During this meeting the tradesmen are blamed for the job not progressing fast enough and are constantly reminded that everyone must push harder to meet their impossible schedules. These job meetings are a lot like town council meetings: 99 percent bullshit.

Politics aside, a big jobsite is like a single cell multiplying to become a complete creation. Usually it turns out to be something beautiful, well planned, and organized. But anybody who understands biochemistry knows when cells split too quickly or out of control they instead become some kind of funky carcinoma. In construction, this is job cancer (or "job carcinoma," as I call it), which is caused mostly by construction managers. The only cure is the lowly tradesmen, or might I say the *highly skilled* tradesmen. It really is amazing how well these chaotic, high pressure jobs turn out with the right skilled labor.

WORKFLOW

The workflow of big jobs occurs in seven stages:

First stage—Start up, gear up, man up

Second stage—Familiarization of job rules, regulations, and procedures

Third stage—Job progresses nicely

Fourth stage—Some bumps and hiccups

Fifth stage—Schedule in jeopardy

Sixth stage—Total chaos, screaming, yelling

Seventh stage—Everybody's happy, and the job's turned out well

Theses stages are often chaotic, but I truly believe that chaos is beneficial to the overall progression of the job. As crazy as it may sound it keeps the workmen on their toes. I have a theory that when an important project must be done quickly, you should give it to the busiest workman in the crew. He may not need or want more work, but he is already in high gear and will get it done. This is a tried and true tactic.

Of course, a construction site is far more than just the chaos of noise, dirt, dust, cement, steel, wire, and machines. It's really all about the individuals that make it happen—the men on the job.

MEN ON THE JOB

On site, many different skills are required, and specialized individuals are needed. This is no place for a jack of all trades. Here, you will find operating engineers, masons (cement workers), iron workers, carpenters, plumbers, fitters, duct workers, electricians, laborers, roofers, plasters, painters, and glazers, all very skilled and very important.

THE ROOFER

If I had to order the trades on a ladder by most desirable, hands down the bottom of the ladder would be roofers. Roofing is the toughest job, period. You can bet that most roofers are ex-convicts, escapees, or even fugitives. They can be scary to look at. A roofer's job is very dangerous and dirty. They work in high places and deal with heavy materials like boiling-hot, sticky tar and smelly shingles. At the end of the day they go home with an Excedrin headache and a sore back. While not my idea of a fun job, it's an important trade that involves a great deal of skill.

THE PLUMBER

Grubby plumbers and fitters (one and the same) go somewhere in the middle of the construction ladder. These are guys who like to show everybody their butt cracks and act like they are drunk or on drugs most of the time. I'm pretty sure that, here in New Jersey, it's a prerequisite for plumbers to regularly drink alcohol. The plumbers' union hall usually knows most of its members before they even are members. Men get their brothers, sons, nephews, nieces, sisters, cousins, and second and third cousins in their unions I'm just kidding, of course. Some of my best friends are plumbers. There is actually a lot to know in the field, and I'm not going to pretend that I know the half of it. Three things plumbers do know without question are:

1. Hot water is on the left
2. Shit runs downhill
3. Don't bite your fingernails

Now, I don't want you to feel sorry for these slugs or think I'm picking on them unfairly. These guys make a lot of money. You see, they aren't your garden-variety house plumbers. They work on brand-new buildings or factories. Everything is new and clean, unlike house plumbers who deal with shitty problems on a daily basis. Union plumbers usually install all-new sewer lines, water lines, heating and air conditioning lines, and sometimes specialty equipment. This so-called "specialty equipment" really gets these butt-crackers scratching their usually balding, lumpy heads.

In all fairness, plumbing is a very technical field, and it does need a few smarts. Some of the tricky vocabulary they throw around when they think other tradesmen might be listening can be impressive, such as "reverse osmosis," "capillary action," or even "cavitation." Yeah, they've got it goin' on.

THE CARPENTER

Moving up one rung of the construction ladder—or maybe down two rungs, I haven't made up my mind yet—let's talk about carpenters. Oh boy, where do I start? Without question, these guys think they are the be-all and end-all: The smartest. The toughest. The only people in the trades that know anything about construction. In fact, they are basically inept and closely related to mollusks. On second consideration, they definitely fall on the second rung from the bottom of the ladder. Carpenters are the bottom-feeders of the construction world. You would think that, being "so smart," these guys would learn how to run an Ace-92 plastic comb through their greasy hair once in a while or to wipe a stick of deodorant on their pits, especially on those sweltering summer days.

Nah, that might kill some of their special brain cells. Admittedly, carpenters do have some skills. They can read a ruler all the way down to one-sixteenth of an inch, occasionally read a basic blueprint, and use power tools, like a circular saw, at least while they still have a thumb and forefinger on the same hand. You don't have to be in the construction industry long to know who the journeyman carpenters are. Just look for the stumps of missing digits on their old, cracked hands.

In the past, for some unexplainable reason, one carpenter on a jobsite was appointed the "general" and was in charge of the other trades. The general foreman of the carpenters (known on the site as the general) represents the general contractor. Ten out of ten times the general is a bonafide dickhead, egotist, and know-it-all. The position should come with a mandatory work uniform with stripes on the sleeves and five fucking stars on the hard hat. This monster, this newly created piece of shit, needs the uniform because, in his mind, he is preparing for the mother of all battles. He is going to war with all the other subcontractors on the job. These dick-head generals think that a monument in their image will be placed in the building's foyer at the end of the job. Idiots! Can you imagine a carpenter as the head of the tradesmen? You might as well appoint a painter; at least when a painter is finished the project, it usually looks good in the end.

THE CONSTRUCTION MANAGER

The practice of appointing a general has gone by the wayside in recent years, and "constructions managers" now fill that role. Let's get one thing straight—these guys aren't even part of the construction ladder, nor would they be allowed on it or near it. They're no-good cocksuckers. Fuck them! They are the newest breed of job-pushers, or supervisors. Construction management companies are popping up all over the place. They send their employees to a four- or six-week course on how to manage a big construction project, and these guys graduate as "experts" in the field of construction. They are young, energetic, and clueless. They know very little about the meat and potatoes of construction, but they're good at using a laptop filled with prewritten programs that print out unrealistic timelines and schedules. They're all soft preppies that you just want to bitch-slap. They walk around the jobsite like they're somebody special in their button-down shirts with the collars sticking out of their V-neck sweaters. One hundred percent jerk-offs. There should be a law that you can't be a project supervisor without at least ten years of hands-on experience. In my opinion they are a waste of time and end up costing individual subcontractors money

THE OPERATING ENGINEER

Operating engineers (OEs) get the site ready for all the other trades to begin. They clear the site and dig for the foundations and underground utilities. The OEs operate tractors, backhoes, cranes, and anything that lifts, digs, or moves dirt. They're really not engineers with a college degree—it's just an important-sounding title. What a racket these guys have. You might think that by working in dirt and mud all day their clothes and boots would be a mess; not so! They are real prima donnas. They don't get out of their temperature-controlled backhoes unless the backhoes have flipped over or are on fire. A poor laborer has to follow the backhoe with his shovel all day, showing the OE exactly where to dig and making sure the OE doesn't dig up any buried electrical lines or existing utilities. You can spot an OE by the way he dresses—super clean and creased blue jeans held up high with a thick belt with a shiny, jumbo buckle; a flannel double-pocket shirt neatly tucked in; and gaudy, super-pointed cowboy boots. Some even wear cowboy hats. You might think they just finished shooting a Marlboro cigarette commercial. OEs aren't really on the construction ladder. They are on a man lift somewhere between the first and second floor.

THE IRON WORKER

We now come to the iron worker: the icon of heavy construction, the cover boy of all the trades. The American flag and the iron worker just go together. Their job is as tough as it gets, and they're as tough as any person can be. They can climb steal I-beams like Spiderman. I don't think I'll make fun of these guys. I always said if our government wants to win a war quickly, send in the iron workers. It's a no-brainer. These guys have nicknames that really fit their appearances: Bear, Gator, Wolf, Snake, Killer, Scar, and Big Foot are just a few that stick in my mind. They are short-timers as far as a job goes. They storm on the site, build a massive steel skeleton, top it off with a Christmas tree and a big American flag, and leave.

That's when all the other trades converge on the scene, and that's when the fun really starts. These guys are near the very top of the construction ladder.

THE MASON

Masons, or brickies, come on the job while the iron workers are still finishing up. The masons pour the cement floors and decks and start laying block walls and partitions. Who in their right mind would want to be one of these block heads? (No pun intended.) Talk about a monotonous job—you lay one block on top of another, on top of another, and so on and so on. It does require the ability to use a level and string line. A good mason will boast about the number of "units" he can set in a day. Units? Is it too hard to say "blocks" or "bricks"? No, but "units" is the lingo, the trade talk. Masons do have to keep the number seven on the top of their mind, because seven courses of blocks are all they are allowed to put up in a day. I'll reveal a little secret: sometimes they cheat and put up nine or ten courses. You've just got to love the robotic mason, the guy with Popeye forearms and vice-grip hands. After all, doesn't everybody have an old uncle who is or used to be a mason? The guys and gals of the mason trade lie in the middle or lower half of the ladder.

THE DUCT WORKER

Oh boy, the "tin knockers"—this crew comes on the jobsite shortly after the floors are poured and before the building has exterior walls. These guys come from the same mold as the butt-crack plumbers. They are downright annoying, and their presence is felt the second they show up. They are the noisiest cock suckers I've ever met, pulling into the jobsite with their antique stake-body truck, reeking of exhaust, surrounded by blue smoke, and loaded to the top with shiny heating and air conditioning ducts of all sizes. This load is pretty fragile, and the shop boys are obviously trained in how to unload it carefully. They gun their stinking-ass truck, pop it in gear, and hope the whole load slips off the truck in one shot. This never works, so they end up unloading the rest by hand. Did I say these guys are annoying? I meant to say they are just downright fucking obnoxious. They earned the name "tin knockers" for a reason. To be a tin knocker you need just two tools, a hammer and a tin snip. These guys are definitely psychotic. All day long they bang on their ducts trying to get them to fit together, very noisy! Needless to say nobody is happy to see the tin knockers show up. Most of these guys come equipped with a stinky fat cigar glued to their bottom lip, which just adds to their overall appeal. If that's not bad enough, they also think everybody should get out of their way, but they soon learn where they fit on the construction ladder: somewhere near the carpenters.

THE SPRINKLER FITTERS

Fire safety is big these days, and any building in New Jersey that's bigger than one story needs a sprinkler system. The plumbers who install these systems really aren't plumbers at all; they are much more than that. They're called "sprinkler fitters," and to be one you must have a strong back; a pair of strong, monkey-wrench arms; and the ability to match up whole numbers. Why whole numbers, you ask? Because the pipes they install throughout the building have numbers on each end. These pipes, which carry water throughout the building in case of a fire, run in sizes from one inch in diameter to over twelve inches. They are precut and numbered to match the connecting pipe. The thought required to perform this process is, let's say, minimal. I remember when I was a little boy my mother bought me a paint-by-number kit of Mona Lisa. It was for ages eight to adult. Need I say more? This group is the third rung from the bottom of the ladder.

THE LABORER

The title says it all. These workers do all the dirty work and demolition. They deal with dust, hazardous particles, dirt, and more dirt. The most important requirement to be a laborer is a strong back, as their main tools are a pick, a shovel, and a broom. I couldn't do, and would never want, this tough job. Most laborers I've met are quite the characters. I think they are on the job mostly for entertainment. They label every worker on the job; no one is spared. Everyone gets a nickname, and it always fits. Lunchtime on a jobsite is segregated by trades. It's an unspoken rule not to intermingle. It's typically only a half-hour, so there's no time to socialize anyway. Every trade picks a room where they will eat lunch every day, and that room is off limits to other trades. There are usually microwaves, refrigerators, tables, and clothing racks—most of the comforts of home. At lunchtime there are lots of smells wafting about, none more interesting than what's drifting out of the laborers' area. The scents are almost gourmet. I'll never forget when one laborer, Willie, kept bugging me to try one of his specialty dishes, which he called "mountain oysters." They were deep fried. It was pretty good . . . until I found out what it really was. You can look it up in the glossary. The laborers deserve a special place on the construction ladder.

THE GLAZER, PAINTER, SPACKLER, AND FINISHER

I will combine the next set of workmen and won't go into much detail. The window workers, technically called "glazers," are about as skilled as sprinkler fitters. Every window comes with an assigned number and goes into its corresponding spot on the building. If a window doesn't fit in the square opening, they keep trying until they find a place itdoes fit. Their finished work puts a nice finish on the building, and they walk off the job like proud peacocks. They act like they just built a nuclear-powered space ship or something. Idiots. The painters, spacklers, and finishers fall into the same group as glazers. They are basically prima donnas. To get them to work, the environmental conditions have to be near perfect; they don't work in the cold, rain, or dusty, dirty places. Seventy-two degrees and dry is all that suits these guys. Yes, they are still classified as construction workers.

THE ELECTRICIAN

Finally, I saved electricians for the topmost rung on the construction ladder. In the construction food chain, the electrician, sometimes called "sparky," is the shark or top-feeder. These men and women must have a broad range of skills. They are able to build anything from scratch, whether from metal, plastic, wood, or cloth. They weld, screw, nail, glue, and are masters of fastening anything to anything else. They hang heavy objects on ceilings or walls and have the confidence to know it won't fall on anyone and kill them. The other trades often come to electricians when they need advice on hanging something of importance or value. Electricians are also masters of rigging, knot-tying, and moving large equipment, not to mention handling high voltages and deadly amperages on a daily basis.

Electrical tasks on large construction jobs are not like your typical house-wiring job. The equipment is large and heavy, and copper feeders and wires can be as thick as your wrist. Wires come in sizes as small as 16 gauges, which is about the thickness of a toothpick, and as large as 1,000,000 million circular mills (MCM), which weighs about 3 pounds per foot. This type of wiring must be pulled through the building in steel conduits with heavy equipment and powerful machines. It takes years to become a master conduit bender/installer. Bending such large and heavy conduits requires special hydraulic equipment to bend, not to mention all of the math needed to fit these conduits into impossibly tight spaces. All wire pulls must be preengineered and set up with great detail. One incorrect calculation could result in an injury or damage to the very expensive wire.

Electricians must have engineering in their blood, as well as a good grasp of advanced math, not just basic algebra. The Pythagorean Theorem is used by electricians every day, and pi is used in most electrical equations or calculations. Of course, when a nosy carpenter or plumber overhears the word "pi" in an electrician's conversation, they want to know if it's apple or pumpkin. Idiots! Other math such as powersfactors, and efficiency formulas are necessary in wiring today's energy-smart buildings. In fact, training schools for electrical workers now offer courses in calculus, nano technology, and rocket science. Just kidding. I think I just got carried away.

But seriously, one of the many competencies of the job description includes building microwave towers. The New Jersey radio tower in West Windsor is 1,050 feet high. When the power and lights were installed they sure didn't ask a carpenter. If you have a fear of climbing or heights, you can't be an electrician. The job also involves installation of complex computer and energy management systems that are in new buildings, along with fire safety systems that are so complex they come with several manuals thousands of pages long. The electrician, by state law, must complete courses, called "ongoing education," every year. The *National Electrical Code* is over one thousand pages long, and a lawyer's mentality is needed to break it down into laymen's terms. The electrician really has a lot on his plate.

The electrician is the most involved trade, and he is the first on the job putting in temporary power and the very last to leave. But the electrician is no different from any other tradesman in how he is stereotyped on the jobsite. You can usually pick him out easily. Typically he will be wearing a Carhartt vest with pockets filled with nice new pencils and a sharpie; fire-proof slacks and shirt; and thick, insulated boots. Of all the tradesmen he will look the most like a poindexter. Beware the nerd; they make great lovers.

THE APPRENTICE

Apprenticeship programs teach young workers how to become qualified journeymen in their respective trades. College isn't for everyone, and construction apprenticeship programs help prepare those who are not collegebound for a rewarding career in the industry. In fact, a skilled tradesman can reach the six-figure mark (with overtime). The schooling is paid for by contractors, so it's free to apprentices. On top of the free education and potentially lucrative payoff, tradesmen receive a pretty good benefits package with pension and medical included. Apprenticeship programs range from two to six years. Carpenters and plumbers have a four-year program, and electricians have a six-year stint. These trades work a forty-hour week and go to night school. The other trades have a similar setup.

It's not easy to get into an apprenticeship program. Electricians usually have two hundred or more applicants every year, and only twelve to eighteen get in, and in my opinion that's too many. That's only 6 to 9 percent of applicants. In the past, if you knew somebody or were the son of somebody already in the local, you were automatically in—but not anymore! New federal guidelines put a stop to this practice, For the record, I didn't know anybody and still don't know how I got in. It was just luck I guess.

The apprentice is a special animal. The first-year apprentice makes a good coffee boy, clean-up person, and rule breaker because he doesn't know the rules. The second-year apprentice is just a better coffee boy or clean-up person. This is the year when they start with excuses, mostly for why they're late. I've heard all the excuses; they're limitless. The one used most often occurs during daylight savings when clocks have to be set ahead or back. Without fail, most second-year apprentices will be one hour late, though the older apprentices are no better. Everyone will say they forgot to reset their clocks or, worse, blame their girlfriends or wives. One excuse I will never forget was from a fourth-year apprentice. He didn't show up for two days—no call, nothing. When he did return he told me his mother's parrot passed away, and his mother needed his support. Then there is the gold-standard excuse: car trouble. Granted, this one can sometimes be true, but often . . . "So you had a flat tire, and it took you three hours to change it? How come your hands are squeaky clean? I'm docking your ass! Get to work!" There are so many more: "A garbage truck was blocking my driveway"; "My dog hid my work boot"; "I couldn't find my work keys"; "My iguana was stuck in the heating vent"; and so on.

As the apprentice progresses we, as journeymen, must have faith they will become qualified journeymen too and keep the industry healthy. Our pensions are depending on them. By the fourth year most apprentices finally get a handle on the basics and are trusted to do more on their own. Sometimes apprentices reach the end of the program but still aren't qualified for the job.

There are many reasons for this. Some just don't have it in them. Others, by no fault of their own, get stuck with a big contractor on a long-term job doing the same repetitive work for five years and don't get a broad view of their trade. Some just have bad attitudes and don't care or learn what they should. When work is slow these are the guys who wonder why they're out of work and sitting on the bench. Some people just don't get it. The apprentice knows just about everything there is to know about anything—just ask him. And to think forty years ago I was one of them.

An apprentice can be good for job morale, and at times it's good to have a younger point of view or idea, even though his idea is not an option. We just make him feel wanted. I've always enjoyed the apprentices I've worked with. Most are eager to learn, and the ones that weren't, well let's just say they didn't work with me long. They were sent back to the hall for a new assignment. You might say if the apprentice isn't cutting it, even after four years, you should just kick him out. Though it's not that easy because the union has too much invested in them, and apprentices pay their dues. Make no mistake about it, apprentices do work very hard, and most want to make a better life for themselves. One thing I always tell my apprentices is that perseverance is the key to success, and when you point a finger at someone three are pointing back at you. Apprentices take a lot of grief from everyone and especially from the contractors. When the owner/contractor shows up at the jobsite he is never happy, swears he is losing his ass on this job, and yells. The apprentice is always the first to get yelled at (and for no reason). The owner rarely yells at a journeyman, and for good reason. Union rules prohibit any conversation with the men on the job. The owner must deal only with the foreman or man in charge.

An apprentice doesn't qualify for a place on the construction ladder. He's on a step ladder for now.

THE PORTABLE SANITATION UNIT

The portable sanitation unit, better known as the john or port-a-pot, has many brand or company names such as Port-A-John, Spot-A-Pot, Mr. Bob, and many more, depending on which part of the country you're in. Among the workmen it's also known as the honey pot, hell's kitchen, detox unit, chemical wishing well, honey well, stink house, and the apprentice's office. Portable sanitation units are a necessary evil on the jobsite. As far as offending the olfactory senses, they aren't so bad in the wintertime. The chemical liquid in the bottom is usually frozen solid, as is the pile mounded on top. It goes without saying that you have to get in and out very quickly because you too could freeze and get frostbite where you really would not want it. But the port-a-pot is at its worst in the dead of summer. When it's ninety-plus degrees outside the pot becomes a bubbling brew, the devils stew, a molten mass of pure stink. It's during this time of year that the most inopportune events occur.

I once saw an apprentice lose the coffee order for twenty men in the honey well. He dropped the list and money in, about sixty dollars, and once something is in there, there is no getting it back. The poor boy came running to me in a panic thinking I would fire him or the men would kick his ass. He soon found out how good natured big burly construction workers can be, although he took quite a ribbing the rest of the job. I've seen guys lose all kind of things in the stinky wishing well. It has claimed false teeth, keys, hearing aids, tools, cell phones, wallets, jewelry, and eyeglasses. Once I saw a work boot in there, which was a first for me. The white-collar worker or office worker couldn't possibly relate to these situations. They get to use modern toilets, the kind that have clean seats, flush, and are indoors.

The size of a job and number of men present determine how many port-a-pots are needed. Even though there might be more than ten, they always seem full or overloaded, and they get cleaned only once a week. Without fail the guy who cleans them will show up at lunchtime when the men are outside under the only shade tree on the site enjoying their meal. The service truck will pull in and back up to the unit and start pumping away. The men eating their lunch always seem to be situated down wind and even if the wind is blowing up wind it will magically shift to blow down wind, that is just the way it always is. If you've never experienced lunchtime downwind of a port-a-pot being emptied, all I can say is there are few words to express how terrible this event is. It's worse than being downwind of Three Mile Island when it had its mishap. Your nose hairs are burned, and your lungs are scarred for life. Oh the humanity! Needless to say lunchtime is quickly over.

So, you ask, what about the poor guy that cleans the pots? Poor guy my ass—he makes six figures a year. He pulls into the jobsite with a $250,000 gleaming Mack truck pumper, which looks like he polished and waxed it all morning just so it looks good at the lunch scene. These trucks are state-of-the-art machines, and their chrome wheels are nicer than Snoop Dog's twenty-two-inch spinners. The operator is so much cleaner than the workers on the jobsite. He climbs out of his rig, slips into his NASA space suit, starts his pumper up, inserts the giant vacuum hose deep into the pot, and in two minutes it's empty, but not until the unbelievable stench has fogged the surrounding area. Next he pulls out a smaller retractable hose from his truck, chemically sprays the whole interior with some sweet smelling liquid, and then refills the well with new green plasma. The operator then effortlessly slips out of his space suit and, before jumping back into his truck, will make sure we are all watching as he lifts one leg and rips a big fart. His job done, he rides off into the sunset.

The freshly cleaned pots aren't too bad; that's the best time to use them. The workers tend to stay in a little longer when they're clean, and that's when the Magic Markers come out. I'm talking about big permanent black markers. The inside walls turn into the job's bulletin board, featuring all the latest job gossip. Nobody is spared, especially not the guys running the job—the supers and foremen get slammed. Every cussword known to man ends up somewhere on the walls. There are poems, catchy phrases, artwork (mostly of sexy women), and some political bullshit. By the end of the job all wall space is taken up. Even the four-inch vent pipe that runs up the back wall is full of graffiti. At this point the job super sets up surveillance to catch the culprit writing shit about him. You wouldn't want your wife or daughters to enter these units. There's a lot of sex talk and graphic pictures, although I find it quite amusing. Some of the graffiti is very funny, and so the port-a-pots are the talk of the job. Someone will say, "Did you see what somebody wrote in pot #3?" And someone will respond, "Yea, but you gotta see what's in #5!" It sounds childish, but the port-a-pots play a big role in job morale. I've been in a lot of sanitation units in my forty years, and there is always something new to make me laugh.

Mysteriously, there's been one phrase that pops up in almost every john I've been in: "CHUDDY BLOWS." I don't know who or what Chuddy is, but boy it must be true. One time I was on vacation in Florida with my family driving a rental car, and I suddenly had to go real bad. I pulled over to a construction site and ran into a port-a-pot just in time. As relief came over my whole body, I looked up at the door in front of me and saw, written in big black magic marker, "CHUDDY BLOWS." Who the hell is this guy? He is an outhouse legend.

Yes, the pots can be quite comical, and I'm glad to say over the years racial comments have virtually disappeared. Less than forty years ago, you would still see offensive graffiti in the johns, from the "N" word to Italian slurs, and Irish and Polish slurs. Occasionally you will see something offensive because there are still a few ignorant people walking around. You just have to deal with it, and they have to live with themselves.

I should also mention that, yes, there are women construction workers too, and by law they get their own port-a-pots on any New Jersey jobsite. I will tell you this: they are no better than the men when it comes to marking up the pots with graffiti, maybe worse! They just use red or blue Sharpies instead of black Magic Markers. I've come to the conclusion that, given the chance, everyone will create graffiti. It's human nature, and everyone's guilty. Anyone in construction knows that there is solidarity among the working men and women, and even though we all make fun of each other, we also respect each other.

So I salute the port-a-pot. What would a jobsite be without one?

WOMEN ON AND OFF THE JOBSITE

The jobsite is usually made up of 99 percent men and very few women. When coffee breaks and lunchtime come around the women go their separate ways.

Often during coffee breaks, men from my local will judge the page-six girl in the local Trenton paper. She is always in a swimsuit, posing in a seductive manner. Of course every guy in the room says he knows her or once went out with her. After the page-six girl is passed around the room she will be either given the nod to be taped to the job box or the thumbs down and thrown in the trash. You would be surprised how critical these guys can be. Very few page-six girls make it to the job box.

If the topic of wives or girlfriends come up during breaks, nobody wants to hear it. It's an unspoken rule not to badmouth family. After all, where would construction workers be if we didn't have the support of our "loving significant other" to go home to after a "fun" day at work in unspeakable conditions and unforgivable weather? It's always a treat to come home to the sweet sound of "Honey how was your day?" Of course, before you can grunt out a sorry-ass "Okay," the honey-do list is flying at you like a swarm of bees. "Before you take your jacket off take the dog for a walk. He hasn't been out all day, and he's about to shit on the carpet. And before you come back in put the garbage cans out front. It's garbage day tomorrow." Yes, Dear. Once that task is done you might think it's clear sailing, but you'd be wrong. Once the dog and garbage are taken care of, you hear, "Dear, don't you think you should shovel the porch and sidewalk while you have your coat on?"

But after all the little chores are done, there is usually a warm dinner waiting, and that does make you glad to be home with the one you love.

The art of communicating will now come into play, and you had better be on your game. It starts off with the usual small talk like "how was your day" stuff. Her day wasn't too easy either. Her morning pedicure didn't go very well. She is sure the Asian girls at the nail salon were making fun of her toes, and the car's dashboard indicator light came on for low tire pressure; I better do something about it soon. To top it off, the Oprah show was depressing. It was about people who had to work for a living. Oh, how I can relate. As dinner winds down and fatigue from hypothermia sets in, she complains about how boring our lives have become and how we don't do fun things anymore. Ugh! The best response I can give her is to say that this spring we will go on a vacation to The Virgin Islands. Little does she know I need it more than she does. Dinner was great. Now it's time for some TV and then off to bed only to start over in the morning. Ah, the life of a construction worker!

THE CONTRACTOR/OWNER

I would classify these people in three groups: cheap, cheaper, and crybabies. Every once in a while you will run into a good one to work for. But for the most part you won't. The contractor will normally hire men out of the local union halls. The hall will send the first man out as a foreman or general foreman depending on how big the job is and how many men the job will need. The foreman or general foreman will meet with the owner of the company and the general contractor running the project to get a feel for what's needed to get this job started. The foreman has the power to hire and fire as necessary. He will go to all the coordination meetings, get familiar with schedules, job rules and regulations, and order the materials and supplies needed to finish the job on time. Owners of a large company may show up at the jobsite once or twice during the duration of the job. If there are money problems you will see them more often.

The owner must put a lot of trust in the man running the job for him. The owner must also have a good working relationship with the union hall because the business manager of the union hall has the final say on whom he sends to run the job. For a year-long project there is usually a two-week setup at the beginning of the job. The foreman will set up his trailers, organize his building plans and documents, survey the jobsite, and figure out the manpower he will need. Most important of all, he will get familiar with the owner's procedures for conducting business, such as what suppliers the owner uses and who the go-to guy is in the office. Whether it's a small or large contractor, the rigmarole is pretty much the same. The only difference is the bigger the contractor, the more bigwigs you have to deal with. I find it's much tougher to deal with a small shop because the owner has his hands in everything, and that ain't good. He will make more mistakes than anybody. I would prefer he just stay in the office, but he will come to the worksite on a daily basis, sometimes several times a day. Without fail he will show up at lunchtime and want to talk about the job with the foreman. Lunch is ruined. If that's not bad enough fifteen minutes before quitting time he will appear to make sure no one is cleaning up early. That's the foreman's job. The small contractor is very predictable. He has to micromanage everything, jobsite and office—and it gets worse. The owner sneaks around the jobsite and then bitches out the foreman about what the workmen are doing or *not* doing. He wants to know why Joe is walking around, even though Joe is going to the material room to get what he needs. Why did Bob take such a long coffee break? Why is Leo using a power saw when he could be using his hacksaw? (Well the power saw is ten times faster, jackass!)

The smaller contractors are always crying poor. I've heard it on every job. How do they lose so much money you ask? Because, in their minds, the men aren't working hard enough or fast enough. Owners have come up to me and told me to fire this guy or that guy for some stupid reason or another. They don't realize how important the workmen may be to the foreman in charge. The owner only seems to see inactivity: workers standing still for a moment, tying their shoes, or lighting a cigarette. What they don't see is how hard the men work the other seven hours and fifty minutes.

When winter comes contractors go to their Caribbean homes with their speedboats and yachts tied up in their lagoons next to their own gas pumps. Meanwhile the slobs that work for their companies are stuck back in New Jersey on a dirty, freezing, dark, steel-framed, dangerous job, suffering from frostbite, slipping and sliding in traffic just to make it to work on time. During wintertime in New Jersey, it's dark in the morning and dark by quitting time. You're lucky if you make it home to thaw out and rest. After a hot dinner at home, it's over, Jack! The warmth of your house settles into your cold bones like a powerful sedative, and you are out like a light. It's not a pretty picture. There you are sitting in your comfy chair, the six o'clock news blasting on the TV, and your mouth wide open making inhuman noises. Owners don't know what it's like to work such a physically demanding job, and in some very adverse conditions to boot. Ultimately, they really are a bunch of crybabies. They cry all the way to the bank after the job is done. All that being said, without the owners we, the blue-collar workers, probably wouldn't have a job.

A JOB WELL DONE

Whatever one may think about construction workers, good, bad, or indifferent, the construction worker couldn't care less. The work can be very rewarding, and at the end of the day, even with dirty clothes and hands, there is a sense of accomplishment. Something was built, something of form and function, and it was built to last a lifetime or more. All construction jobs come to an end sooner or later, and when that time comes, all the workers will reflect back on the many challenges that they overcame, such as the terrible weather conditions, hectic schedules, and physical hardships. But they will also remember the relationships made with other tradesmen, all the laughs and jokes played on each other. They will undoubtedly say to themselves that it was a great job. With the job completed, they gather their tools from the job box for the last time, jump into their American-made pickup trucks, and head to the union hall to sign in for new assignments. For each construction worker there will come a time, when driving down Route One with his family in his car, when he will look over at the huge buildings on either side and proudly say to his wife and kids, "I helped build those."

GLOSSARY

These are terms or phrases that one would hear at a jobsite on a daily basis. Some can be very offensive and are not for the faint hearted. Every construction worker in this country has used or heard what follows.

anteater—uncircumcised pud

asshole—general term for boss

ass wipe—boss's patsy

bbs—small breasts as (bbs on a board)

bacon strip—neatly trimmed pussy hair

ball buster—anybody who's unhappy

ball sack—testicles

ball washer—the boss's yes man

balloon knot—butt hole

bat wings—a sweaty ball sack that's stuck to each thigh (a summer condition)

bearded clam—unwaxed pussy

bitches—another word for hot women

blow job—not a windy jobsite

bone smuggler—a gay thing

boner—a stiff pud

booty—a woman's ass

brown boulevard—a gay highway to ecstasy brown eye—but hole

brownnoser—boss's pet

bush—pussy waxed or unwaxed

but lick—boss's pet butt sniffer—a homosexual butter face—everything but her face

C-word—cunt or nasty witch

caboose—rear end

camel toe—normal sized clitoris

chicken choker—a real jerk-off

chocolate starfish—butthole

clitoris—camel toe, moose knuckle, man in the boat, hot spot, clit

company man—a boot licker, ass kisser, brown noser, yes man, ect.

coochie mamma—nice looking middle-aged lady

dick—prick

dick licker—boss's patsy

dick wad—an unpopular workman

dirty dog—a pooch that needs a bath

dunage—wood pile

down spouts—saggy boobs

deuce—#2

fat chicks—roll them in flower to find the wet spot

f.m.—fat momma

foreskin—a foreman

fried eggs—a chick with a flat chest

fuck face—a foreman

fuck head—a foreman

fucker—anybody you don't like

fumunda—something that comes from underneath

gams—woman's upper legs

gap—the space between a woman's upper thighs

gash—vagina

Hershey highway—the poop shoot

Hershey squirts—the runs

highway—what you're on if you don't like the boss's way

honey pot—port-a-pot

honeywell—port-a-pot

hooters—a bar/ restaurant

idiot—a general foreman

jerk-off—a stupid worker

jizzum—a release of tension

john—a port a pot

jugs—big ones

jumbos—bigger ones

knob polisher—jerk-off

landing strip—waxed vagina

lizzard—a pud

lunch—the quickest half hour of the day

mammer jammers—nice boobs

mary fister and her five sisters—a hand job

melons—round boobs

moon—the thing women howl at when its full

moose knuckle—fat chick's clit

morning wood—the first hard on of the day

mountain oysters—pigs balls, deep fried or baked

naughty—a good woman

oops—a mistake

orifice—any hole

outhouse—rest room

pee hole pounder—a jerk-off

phantom shit—no visible turd in bowl after movement

phantom wipe—t.p. comes up clean

pinch a loaf—#2

plow mule—big bottom girl

pocket pool—what a nervous apprentice does when he get hollered at

pole smoker—someone who gives blow jobs

port-a-pot—a necessary evil

prick—the boss

pud—part of a man's anatomy

puntang—trim

pussy—a small cat

rosy palm—hand job

rosy rectum—a sore ass after the boss chews you out

shmertz—something messy

shop rocket—a company man who run around just to look good

skid marks—shit stains in the bowl or undies

snapper—pussy

snatch—pussy

spooning—eating soup

stink house—port-a-pot

taint—space between balls and anus

thongs—butt floss

tits—one of man's delights

trouser trout—another word for schlong

waxed—the way it should be

wazoo—the big one

wood—what a guy gets in the morning

zoo—most construction